Dinosaur Adult Coloring Book

Dinosaur Coloring Book
A Adult Coloring Book containing Dinosaur images filled with beautiful and stress relieving patterns.

by The Coloring Book People

ISBN-13: 978-1530901784

ISBN-10: 1530901782

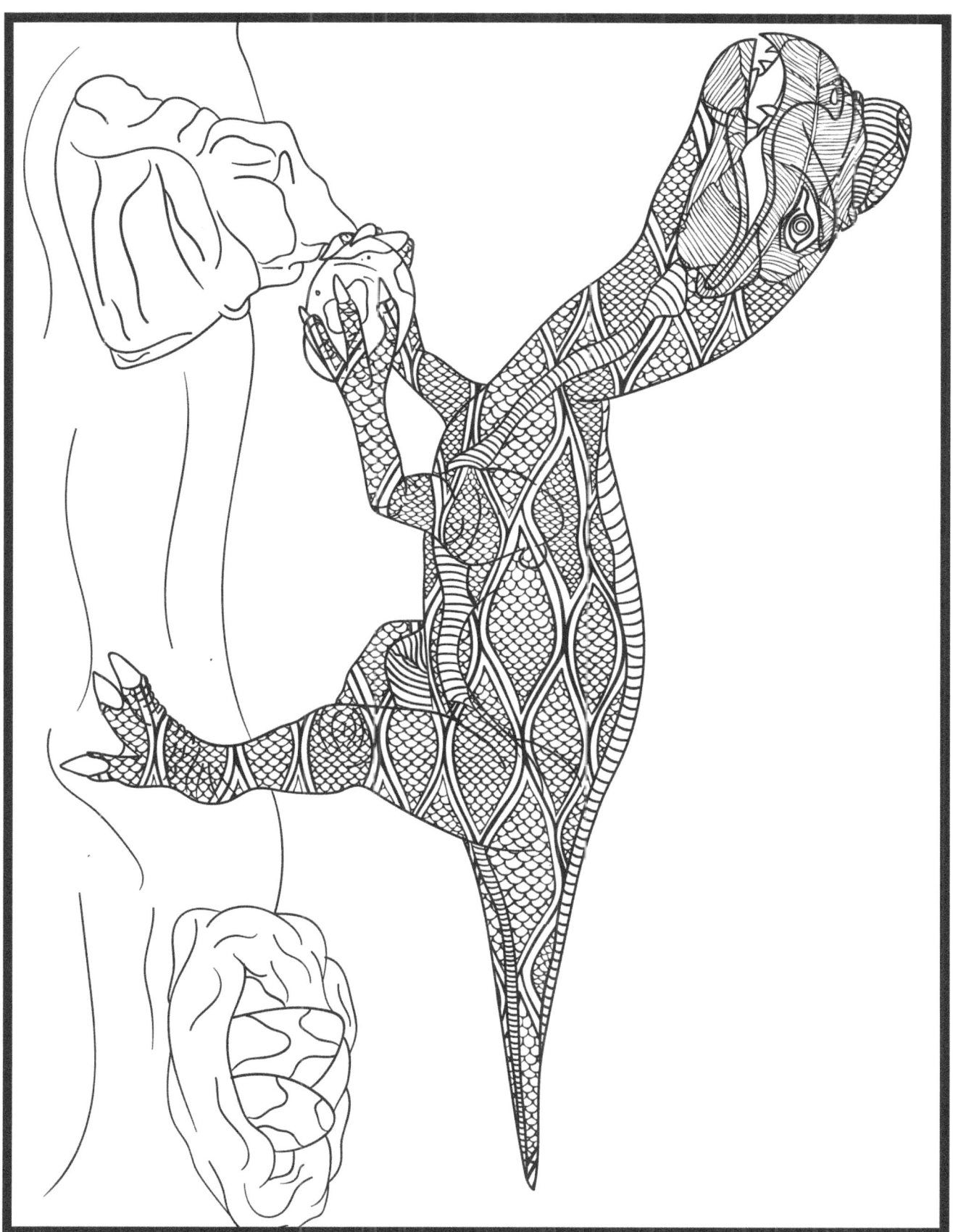

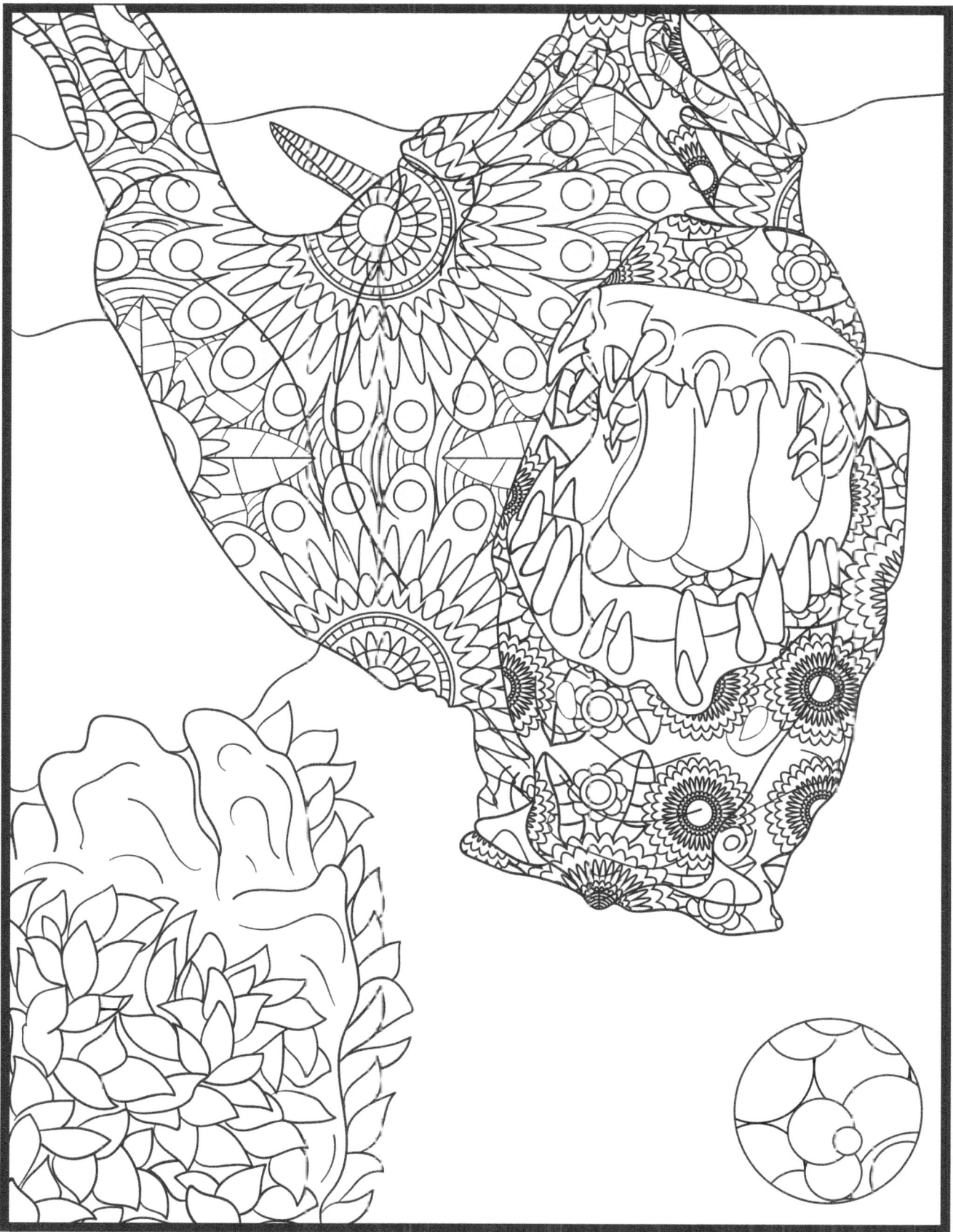

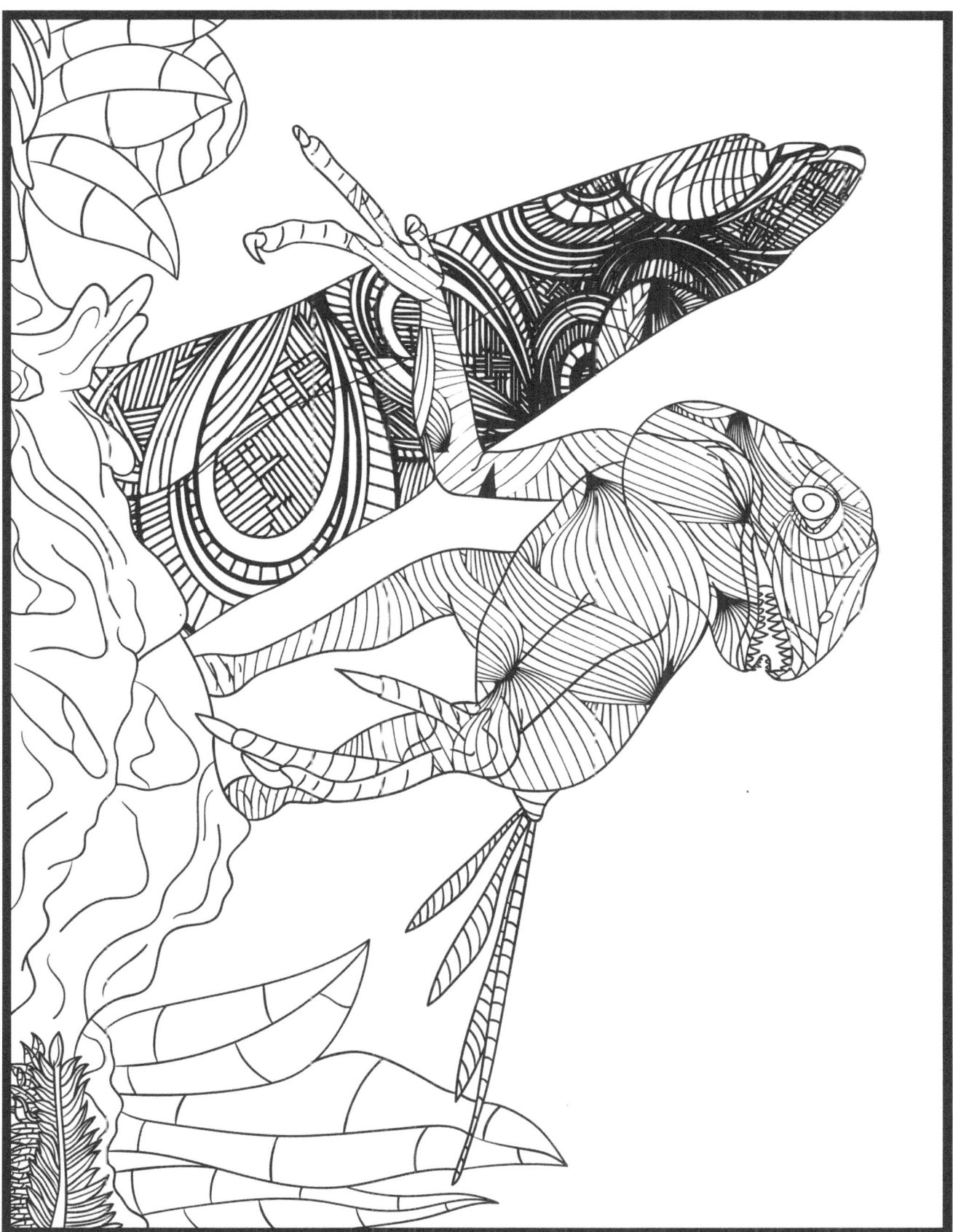

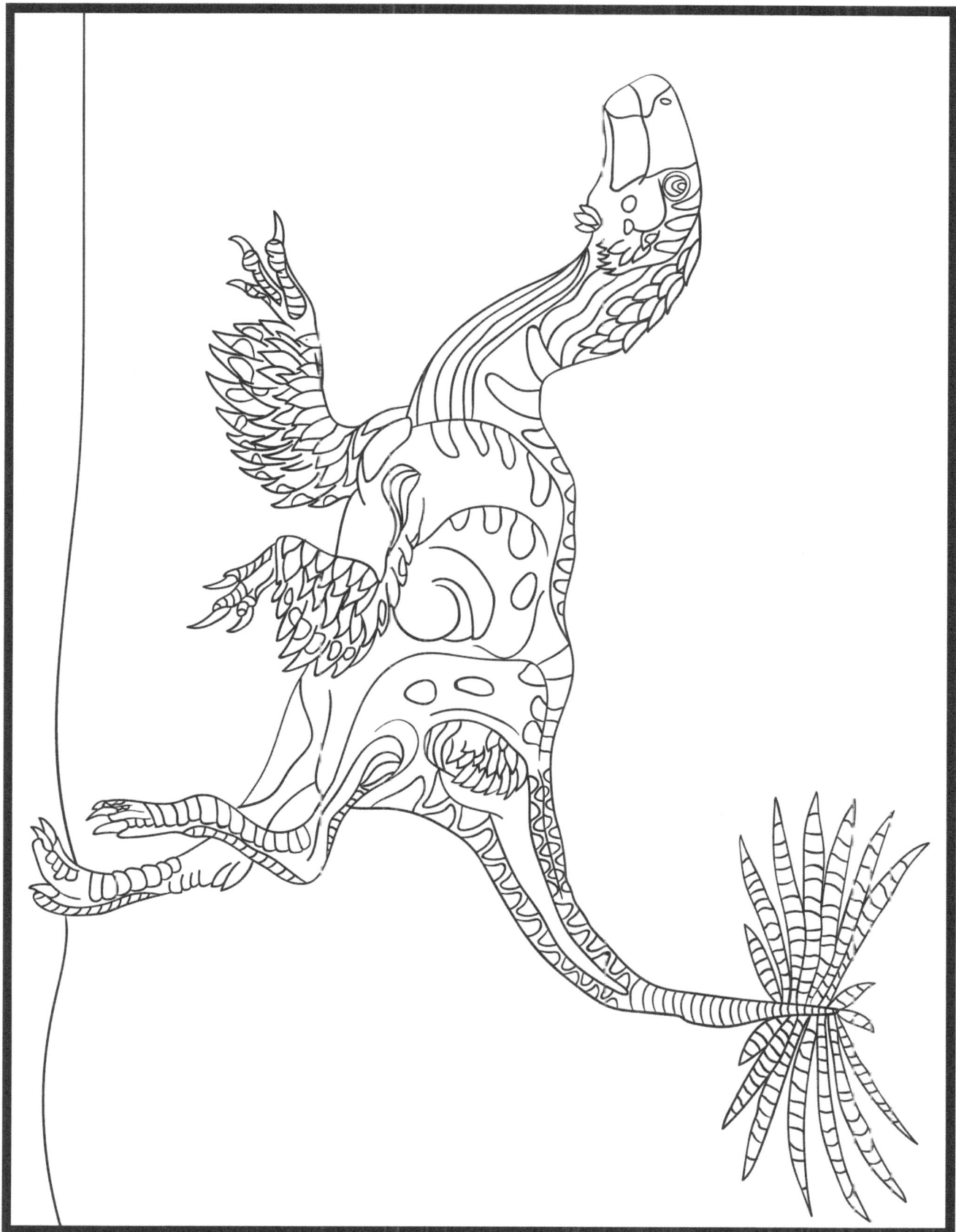

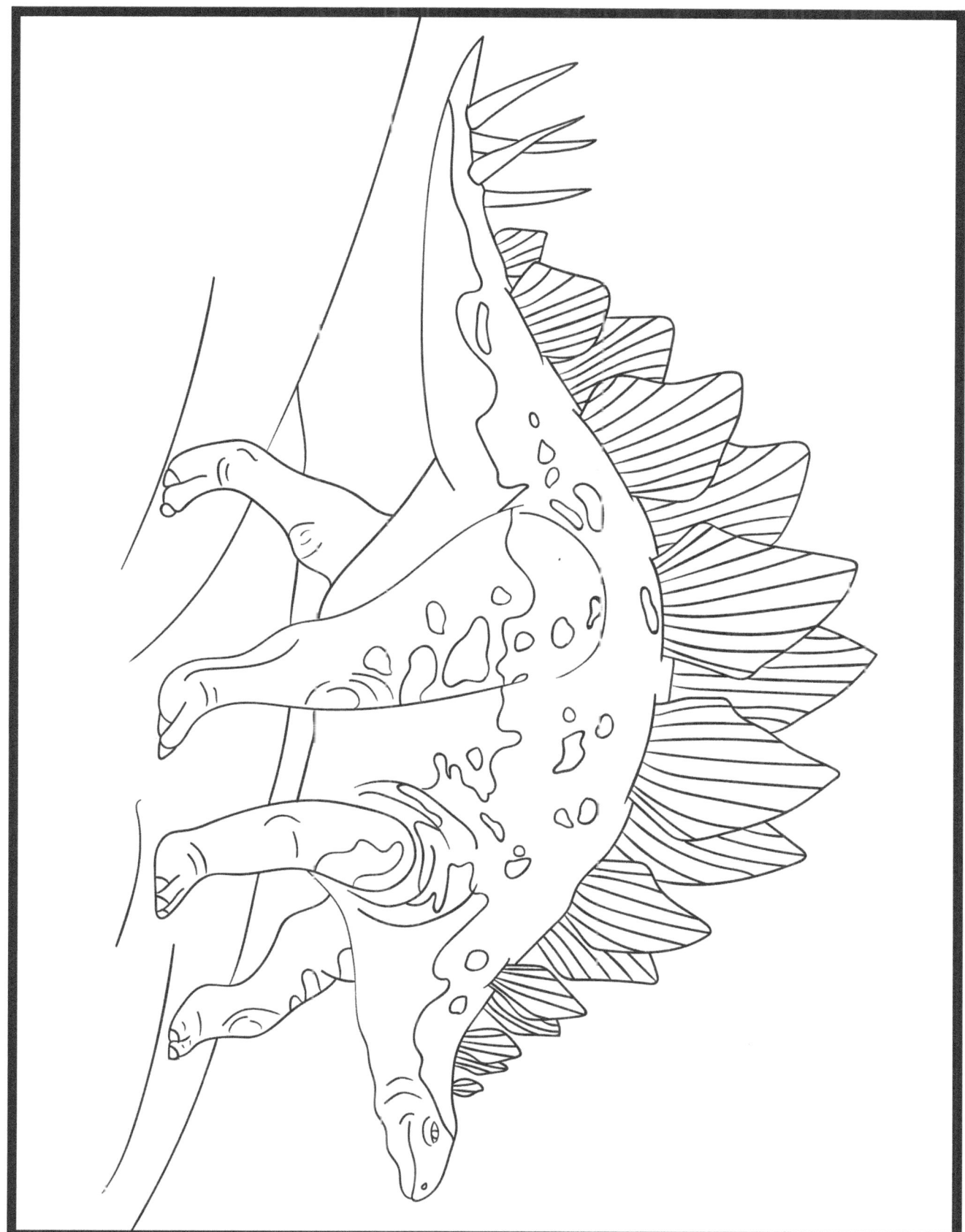

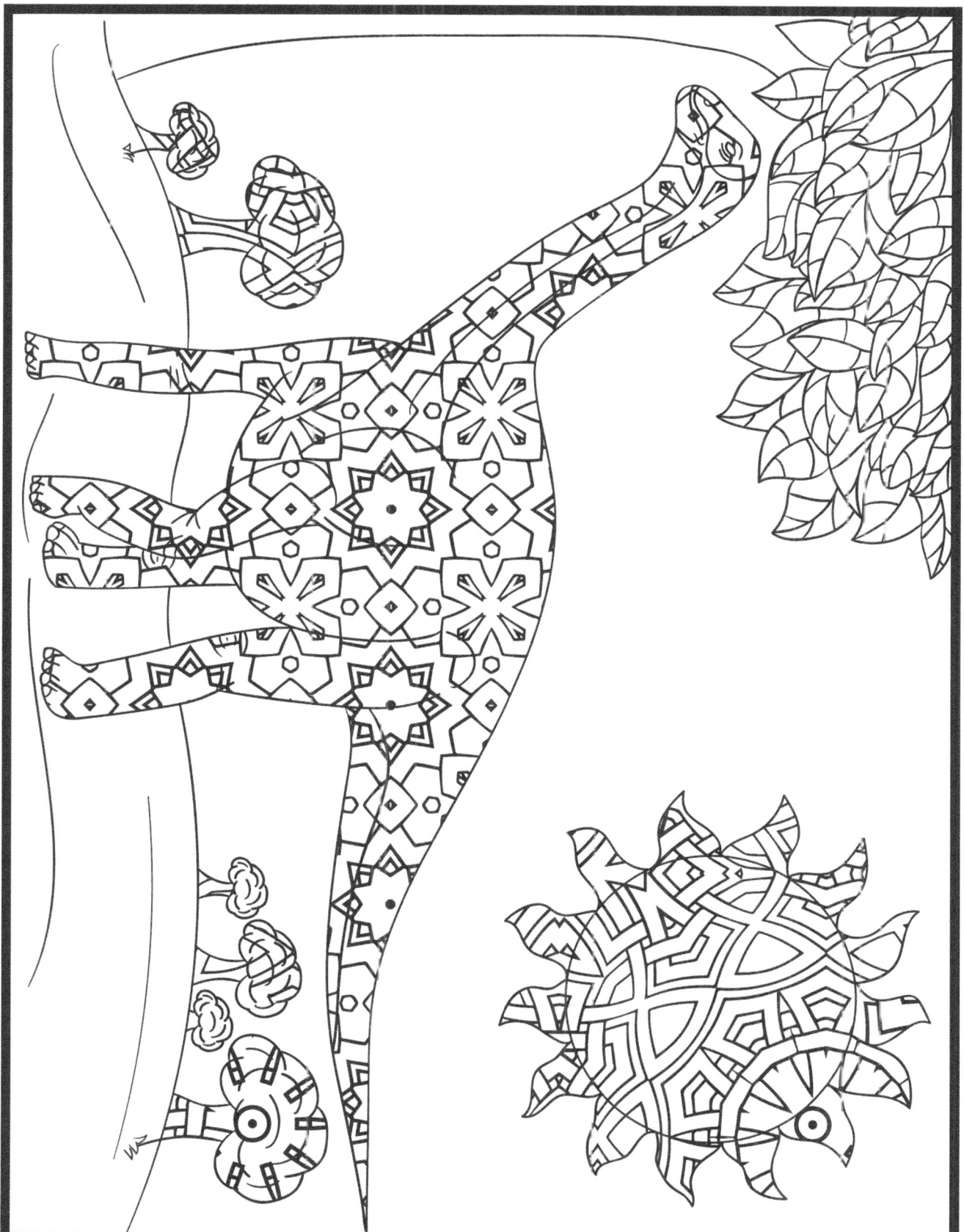

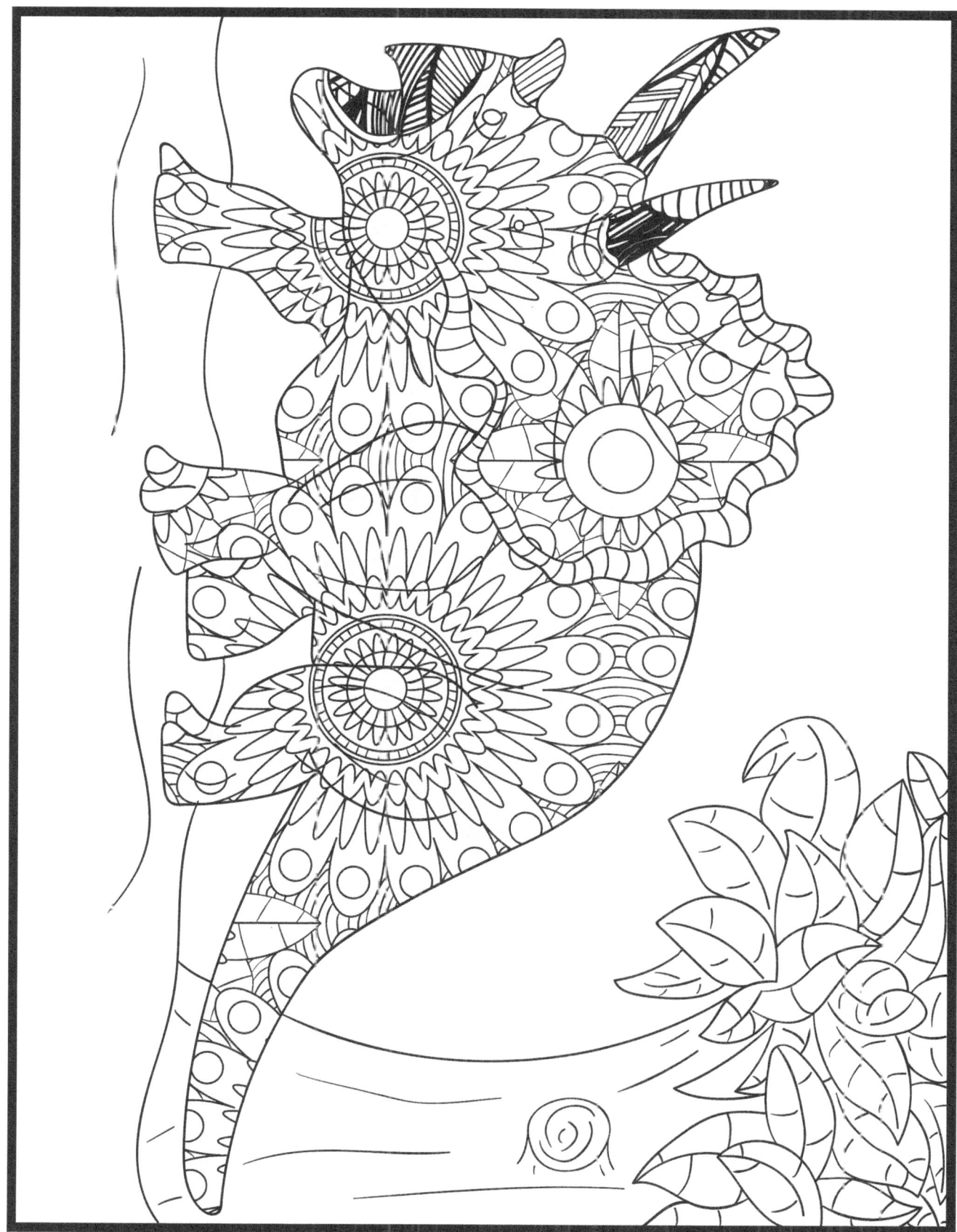

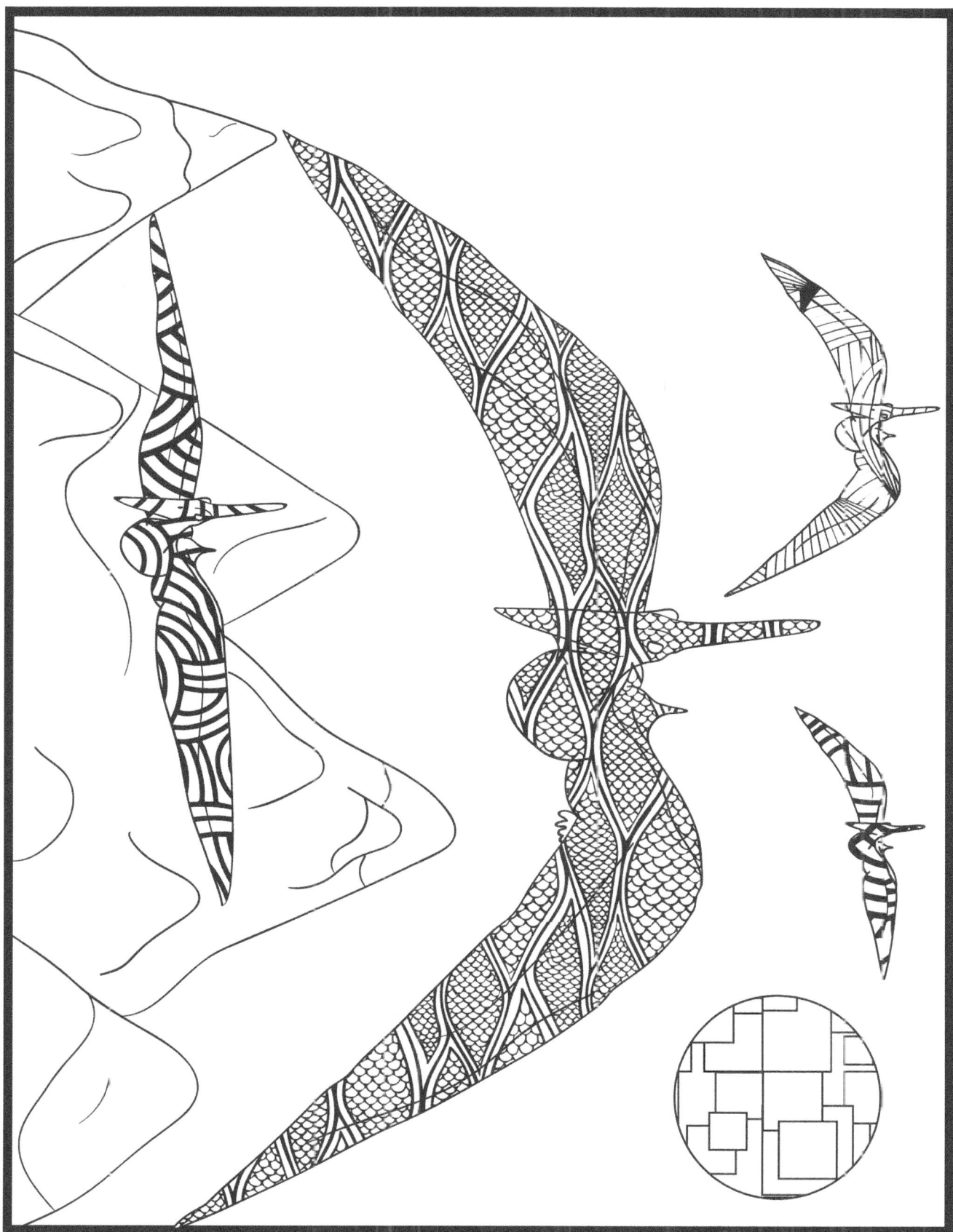

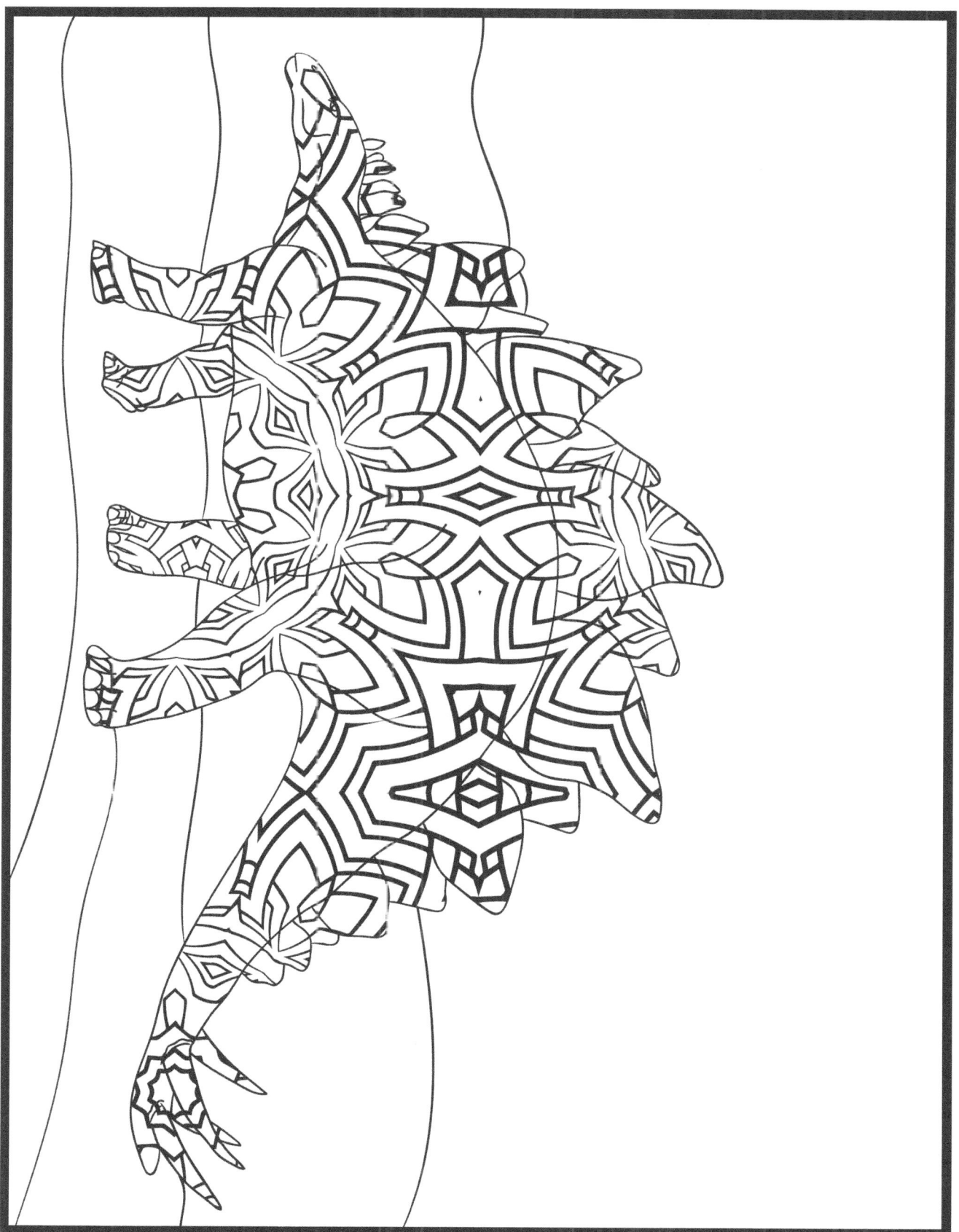

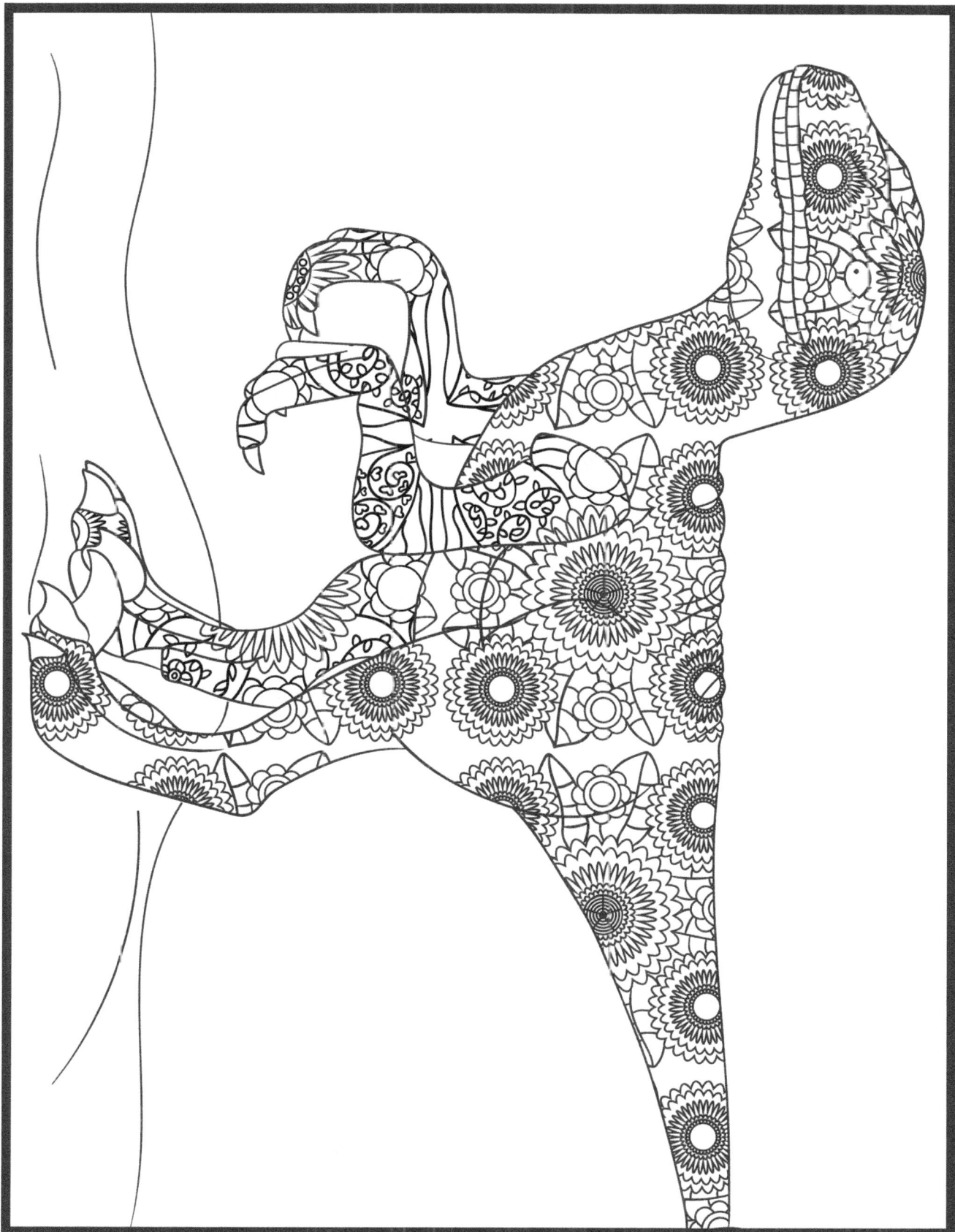

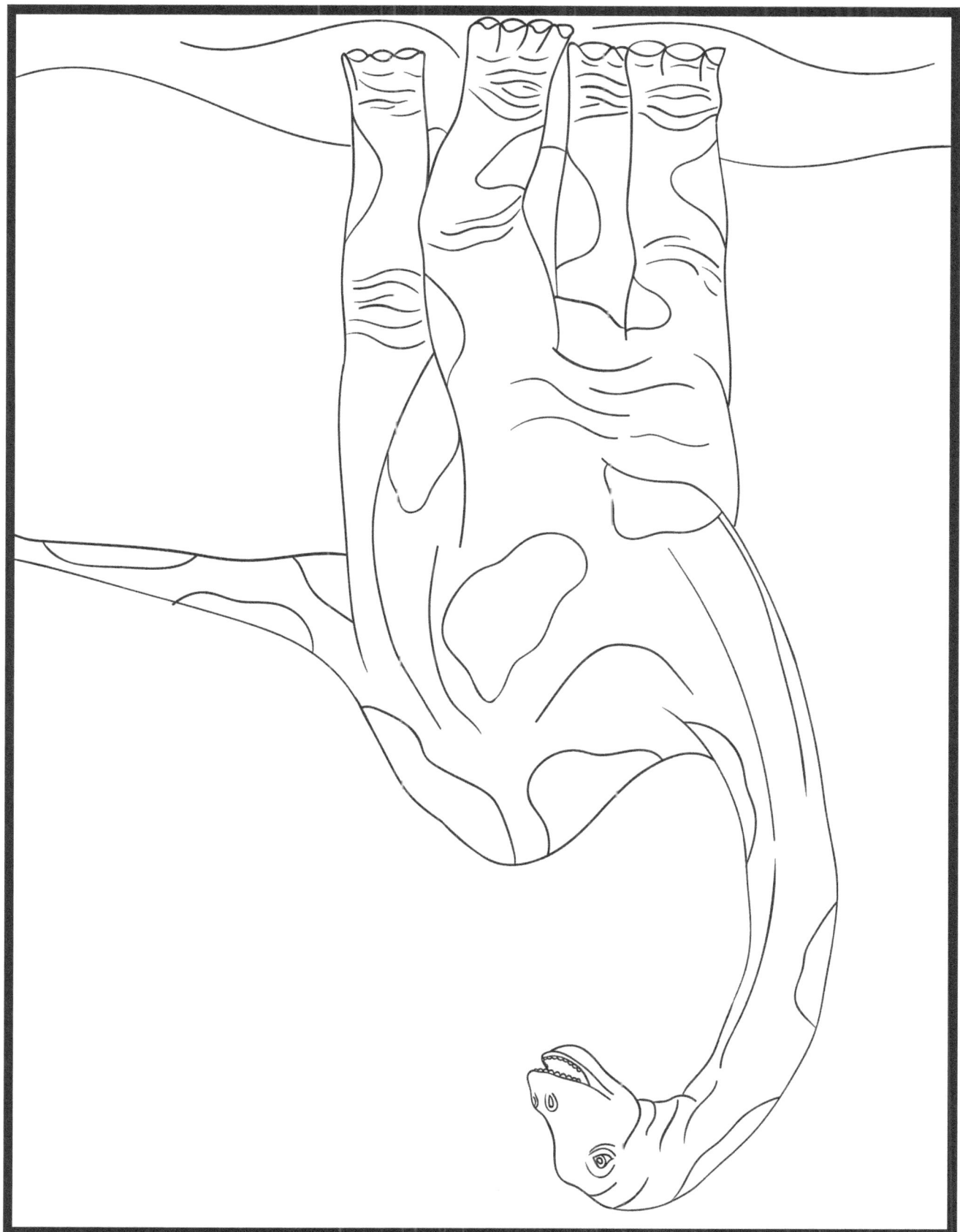

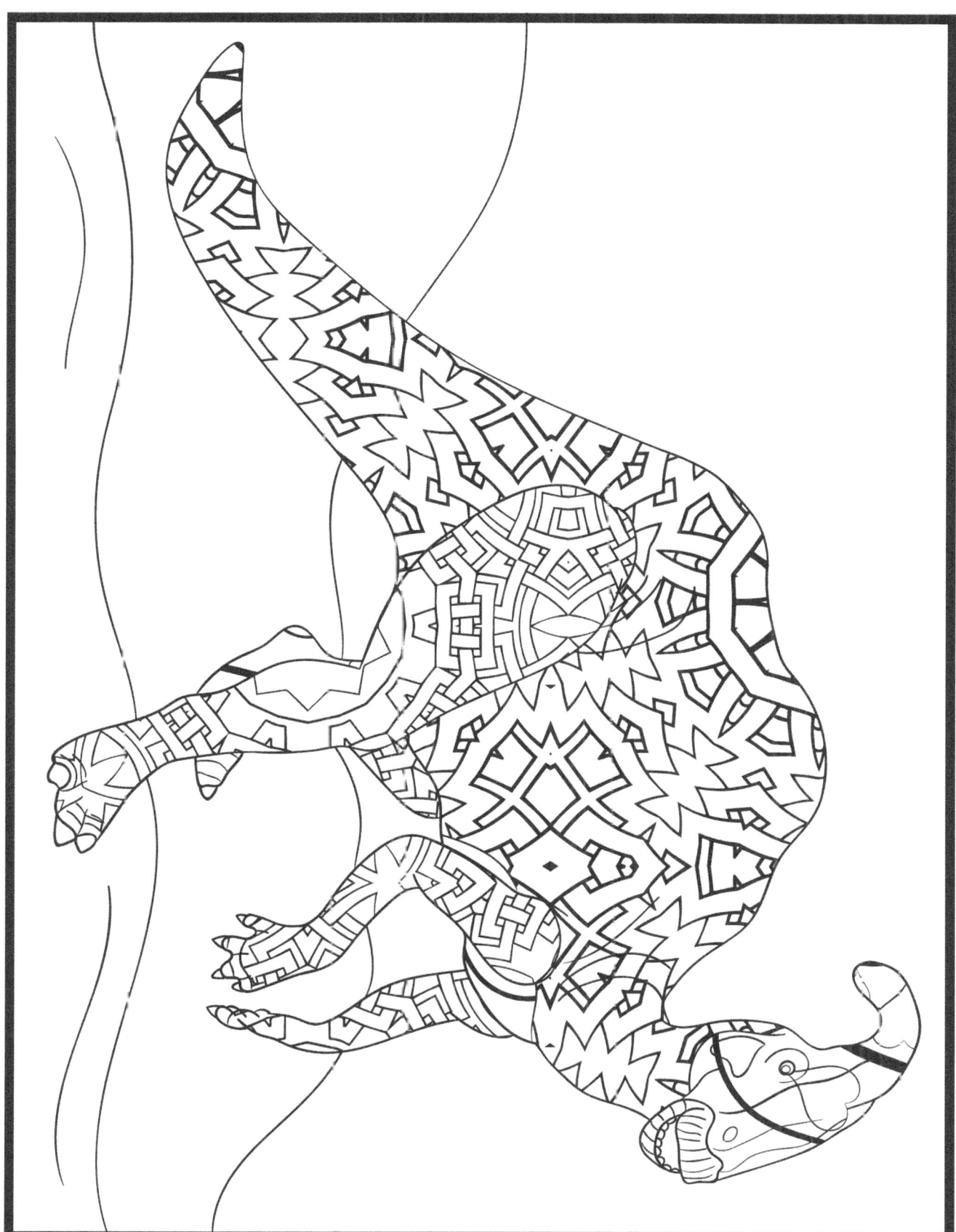

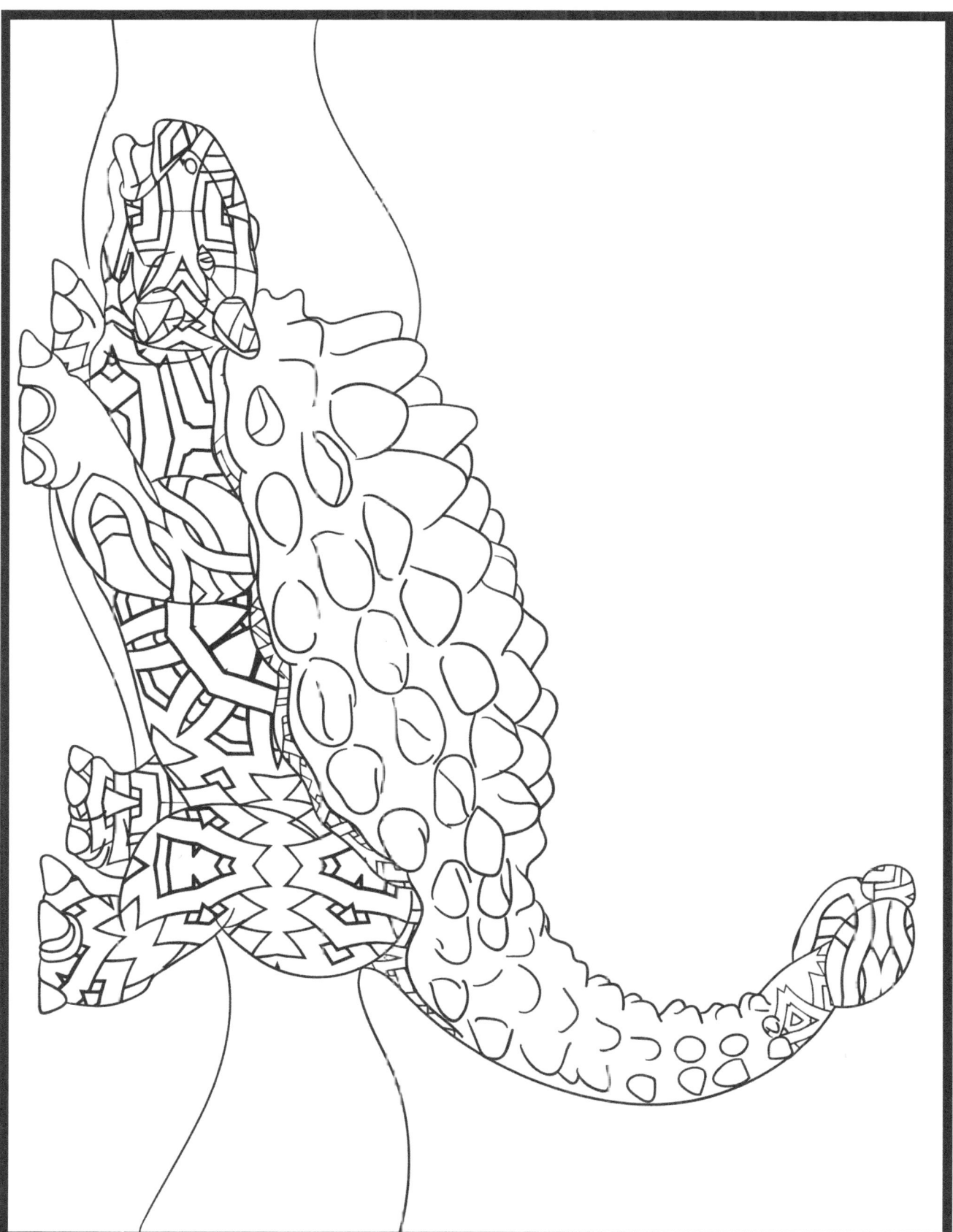

www.ingramcontent.com/pod-product-compliance
Lightning Source LLC
Chambersburg PA
CBHW080611190526
45169CB00007B/2969